# German Infantry Carts, Army Field Wagons, and Army Sleds 1900-1945

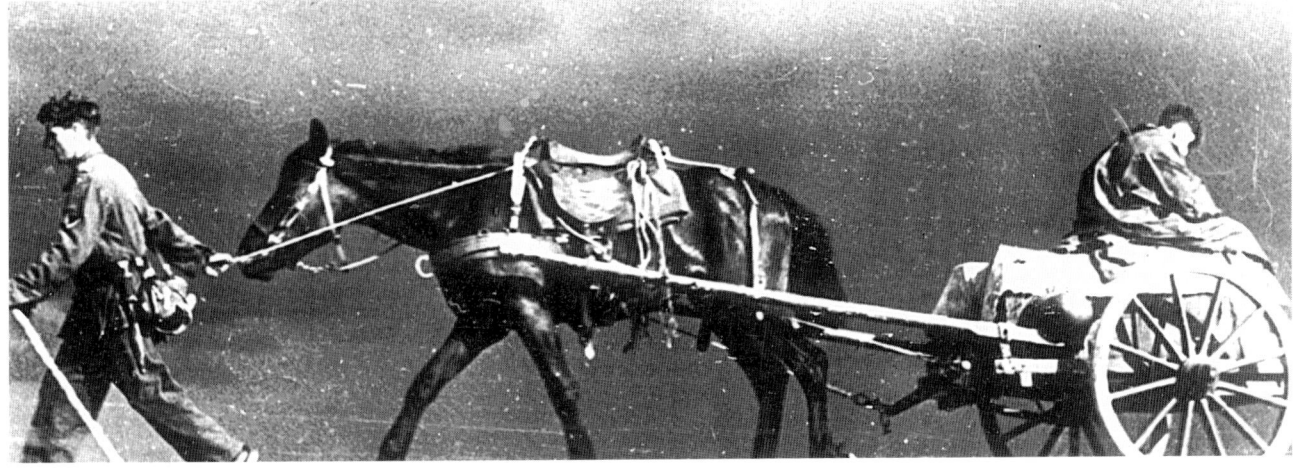

*Not a pleasant coach ride. The war in the east placed particularly heavy burdens on the Wehrmacht's horsedrawn units, in terms of the physical and psychological endurance of man and beast. Countless cadavers of draught and riding horses could be seen on the battlefields and the borders of march routes.*

Wolfgang Fleischer

**Schiffer Military History**
Atglen, PA

# Bibliography

Military Intermediate Archives, Potsdam
WF 03/3207, WF 03/7349, WF 03/13560, WF 03/17366, WF 03/17377, WF 03/31/856, WF 03/35222;

D.V.E. No. 321 Service manual for baggage and ammunition columns and trains, Berlin 1908;

H.Dv. 11/2 The Troop Horse, Berlin 1938;

H.Dv. 112 The Army Wagon for heavy grenade launchers, Berlin 1942;

H.Dv. 465/1 Wagon Driving, Berlin 1936;

H.Dv. 476/1 General Military Equipment Vehicles, Berlin 1936 and 1942;

D 193/1 The Infantry Carts, Berlin 1941;

D 575/2 Loading plans for equipment wagons (Hf.1)..., Berlin 1940;

D 829 Loading plans for the light radio wagon (Nf.3), Berlin 1941, Winter Road Service, Berlin 1943, and many others

## Periodicals

Artilleristische Rundschau, Die Wehrmacht, Signal, Militaerwochenblatt

## Photo Credits

Thiede (16), Fleischer (63), MHM (15)

*The four-horse hitch of an Af. 12 observation wagon pulling the wagon's front section at a trot.*

Translated from the German by Ed Force

Printed in China.
ISBN: 0-7643-1273-1

This book was originally published under the title,
*Waffen Arsenal-Deutsche Infanteriekarren, Heeresfeldwagen und Heeresschlitten 1900-1945*
by Podzun-Pallas Verlag, GmbH

We are interested in hearing from authors with book ideas on related topics.

Published by Schiffer Publishing Ltd.
4880 Lower Valley Road
Atglen, PA 19310
Phone: (610) 593-1777
FAX: (610) 593-2002
E-mail: Schifferbk@aol.com.
Visit our web site at: www.schifferbooks.com
Please write for a free catalog.
This book may be purchased from the publisher.
Please include $3.95 postage.
Try your bookstore first.

In Europe, Schiffer books are distributed by:
Bushwood Books
6 Marksbury Ave.
Kew Gardens
Surrey TW9 4JF
England
Phone: 44 (0) 20 8392-8585
FAX: 44 (0) 20 8392-9876
E-mail: Bushwd@aol.com.
Free postage in the UK. Europe: air mail at cost.
Try your bookstore first.

# The German Infantry Carts and Army Field Wagons
# 1900-1945

*A four-horse Field Wagon 95 used to haul provisions by the Royal Saxon 2nd Train Unit No. 19.*

Until the invention of the gasoline engine and the coming of the automobile as a military means of traction and transport, the picture of the fighting forces was marked by vehicles moved by the muscle power of draft horses, and in some cases by that of men. In Germany the field vehicles, which was the general term for infantry carts and army field wagons, played an important role in the supplying and mobility of the troops even to the end of World War II. Of the more than 300 divisions of German forces in 1944, only comparatively few were fully motorized or armored divisions. In far more than 200 divisions, guns, limbers, and field wagons drawn by horses were part of the scene. The field vehicles, with few exceptions, resembled the wagons that had been introduced into the German Army around the turn of the century.

Before World War I, the Germans had followed the path of specialization in the logistical reinforcing of troop action. That meant that the army's infantry, artillery, foot artillery, cavalry, and engineers used vehicles made to suit their needs in their baggage and ammunition columns. Thus, their vehicles were an extremely varied lot.

In general, the field vehicles were divided into:
    1. Limbers,
    2. Wagons (with four wheels),
and further, according to the material used,
    1. Steel vehicles,
    2. Iron vehicles, and
    3 . Wooden vehicles.

Further differentiation depended on their use and equipment: ammunition, artillery shell, supply, provision, fodder and baggage wagons, field smithies, and field kitchens. And, finally, the wagons were classified according to weapon and troop types. Some field vehicles, such as the light box wagon, served as provision or baggage wagons in various service arms. Others, though, like the ammunition wagons of the foot artillery or the gas wagons of the field airship units, were designed only for one very specific purpose.

The most important material of which field vehicles were made was wood. White beech in particular was used; the screws, nails, and other hardware were made of iron or steel. Shafts, hitches, and other parts were also made of wood. Only later was steel used. Less wood was used for limbers, ammunition wagons, and field kitchens. Horsedrawn columns formed the last link in the system of supplying the troops. Their action could not be limited to roads; they had to be able to follow the troops on rough paths and, if necessary, through the fields and bomb craters. Thus, along with the requirement of the lowest possible weight (they had to carry a load of up to 60% of their empty weight) there were further general requirements. These included security against tipping and ease of steering. In some vehicles, up to 80- or 90-degree turns of the shaft from the center line were possible. Further requirements concerned ease of driving and turning. For that reason, infantry bullet wagons and artillery ammunition and observation wagons were formed of front and rear sections.

3

Above and below: A prescribed six-horse hitch with description of the harness and saddles that were used in the German Army from 1895 on.

The following general factors influenced off-road capability and maneuverability:

1. The arrangement of moving parts, such as wheels and axles. The larger the wheels were, the less pulling power was needed to move the vehicles. In Germany the wheel height was 1400 mm for the field artillery and 1550 mm for the foot artillery. The tire width was between 65 and 76 mm.

2. Stability, holding the chosen direction of travel.

3. Steerability; here, as already noted, the greatest possible deviation of the shaft from the center line was strvien for.

4. Turning ability between the front and rear axles.

5. Vehicle length.

6. Load balance between the front and rear axles.

7. The system of attaching the motive power, meaning shafts, hitches, etc.

The types of hitches used on field vehicles varied according to their forms and uses. Basically, low width and depth were sought, so as to utilize the animals' pulling power better without having to reduce mobility and maneuverability. It was also important to be able to hitch and unhitch quickly in certain combat situations. Aside from the one-horse carts, field vehicles were drawn by two to six horses, divided into side-by-side pairs. The pole horses were at the shafts. They were differentiated into front and middle horses. The horses to the left were called saddle horses, for the drivers could ride on them. To the right the hand horses were hitched.

A six-horse hitch was composed as follows: The pole horses pulled on a rear hitch and carried the shaft by means of a steering chain (balance system). The middle horses pulled on the front hitch, which was attached to the end of the shaft. Like the middle horses, the front horses were attached to the front hitch by long lines.

Naturally, there were certain rules for selecting the horses. The front and pole horses had to be especially strong, and the front horses also had to be experienced. The less usable horses were utilized as middle horses.

The average pulling power of one horse was between 400 and 600 kilograms, with higher performance attainable on good roads. When driven from the saddle, naturally the weight of the rider carried by the horse had to be deducted, for it affected the horse's pulling power.

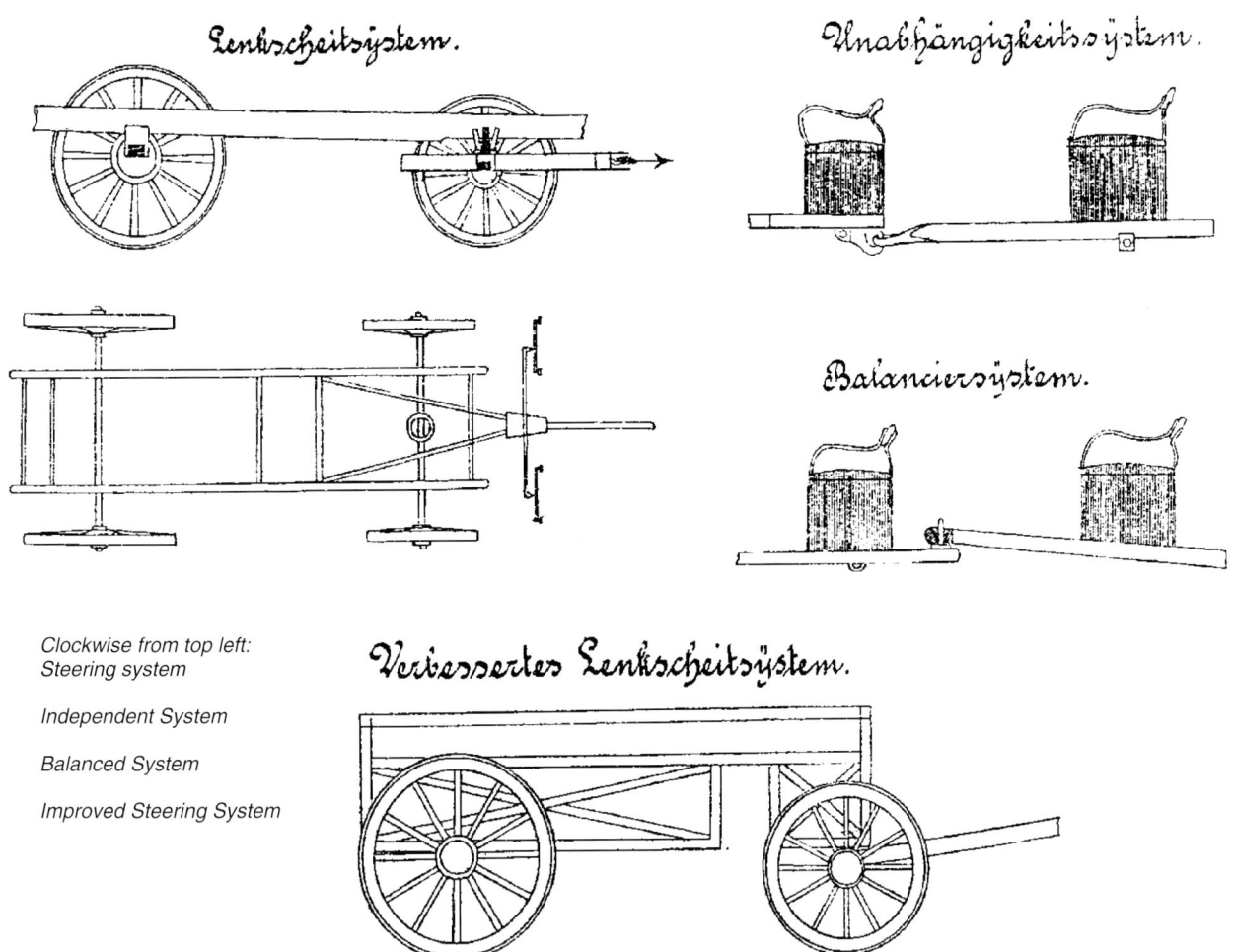

*Clockwise from top left:*
*Steering system*

*Independent System*

*Balanced System*

*Improved Steering System*

A horse's ability to work could be estimated at some four to eight hours per day. It was possible to march 30 kilometers in one day.

At the end of the 19th century the army possessed a scarcely comprehensible variety of obsolete, variously usable and newer vehicles. Some field vehicles remained in troops' service for a remarkably long time in the peacetime years from 1871 to 1914, although armaments had been changed several times during that time. For example, the c/59 ammunition and baggage wagons were only eliminated around the turn of the century.

Naturally, the variety of field vehicles encumbered the training and functions of the troops. For that reason, before World War I there were already efforts to create certain standard models. A pathfinding design in this realm was the c/95 field wagon. It could be used equally well to transport ammunition, baggage, provisions, or fodder for the horses, as well as being used for writing rooms and other purposes.

During the course of the war, many changes were made in the army's organizational structure. It became necessary to eliminate the vast array of specializing in supplying the troops advocated before the war. A decrease in the variety of field-vehicle types was also required. Material shortages caused by the war had a considerable effect on production. As typical vehicles with almost universal possible uses, the Small Field Wagon 16 (Type 05) and the Heavy Field Wagon

05 (also Heavy Provision Wagon 05) took shape. The latter was used with slight modifications as an ammunition or provision wagon in various service arms. There was also a variation of the Small Field Wagon 16 with a sprung wagon body. According to the following instructions, the wagons were painted in field gray, sometimes changed during the war to a multicolored camouflage paint job.

The introduction of new weapons during the war also required the use of new field vehicles by the troops. The minelaying troops, for example, used special handcarts to transport mines. For infantry guns the uniform limber of the field artillery was used. Such special designs as field radio stations and single-axle field wagons (with detachable telescopic masts), two-axle delousing stations, searchlight wagons, and makeshift types for the 3.7 cm revolver cannon when used as an anti-aircraft gun on an adapted field wagon are also of interest. Mobile field kitchens had already reached the troops before the war.

According to the field veterinary report of the German Army, 1,236,000 horses had been used in World War I. The losses were later estimated at 68%. For example, in the winter of 1915 over 800 dead horses were counted on a 15-kilometer road in Serbia alone. During the war, 558,954 horses were treated for complete exhaustion. In addition, the aforementioned report indicates, 405,101 horses survived the war with wounds.

5

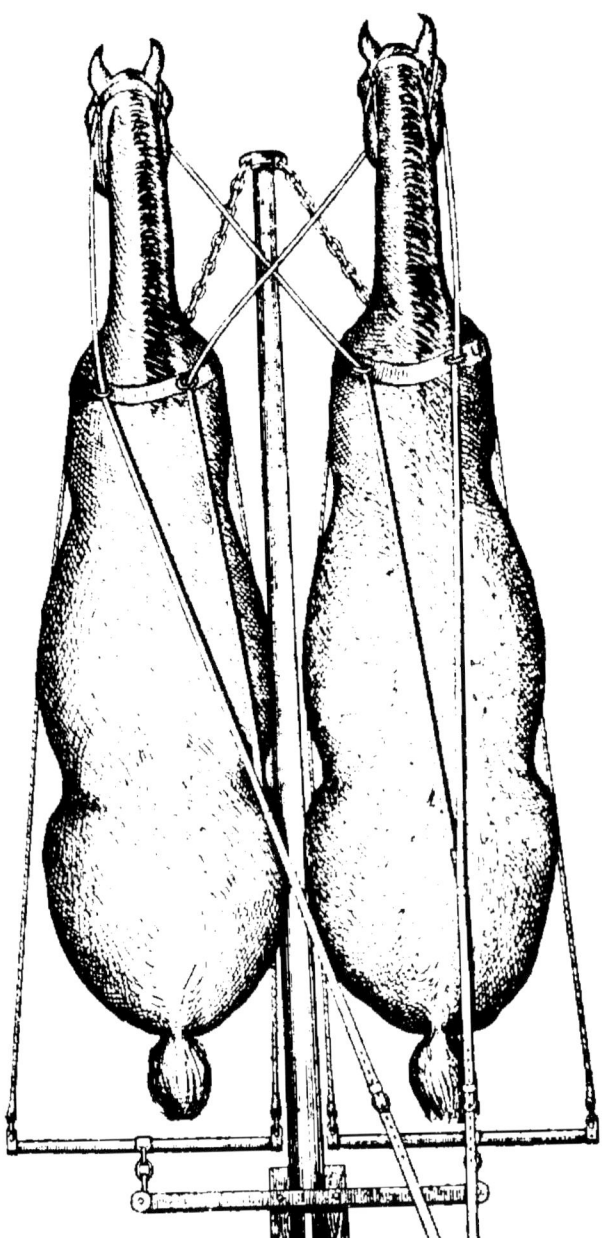

Left: This drawing from the driver's instruction book shows how the horses were properly harnessed.

Below: The two-horse Company Bullet Wagon c/87. It was used by the combat baggage trains of infantry companies and battalions. For every rifleman in the company there were 50 bullets carried on these wagons. The c/87 model replaced the c/59 and c/74 Company Bullet Wagons in the troops.

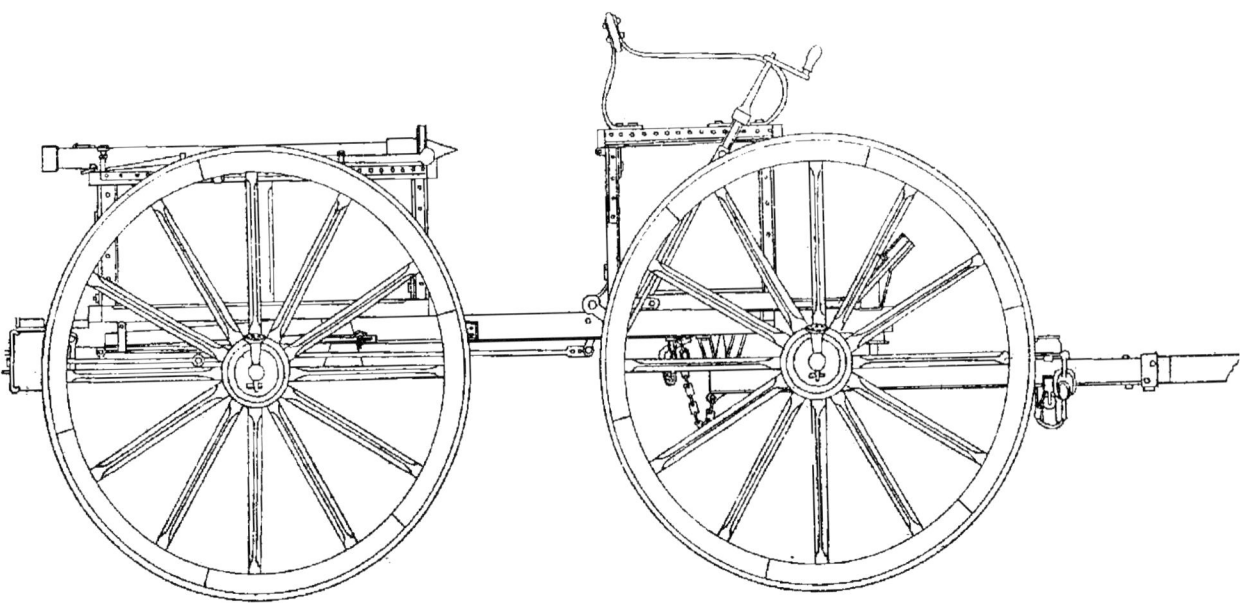

The Baggage Wagon c/87 (later Baggage Wagon 87) was used by, among others, the batteries of the heavy artillery. It weighed 0.85 tons.

A contemporary instruction drawing portraying Baggage Wagon c/87 (also c/1887), from the collections of the Military History Museum of the Bundeswehr in Dresden. The wagon was also used by medical companies. Its wheelbase was 1665 mm, its track 1530 mm.

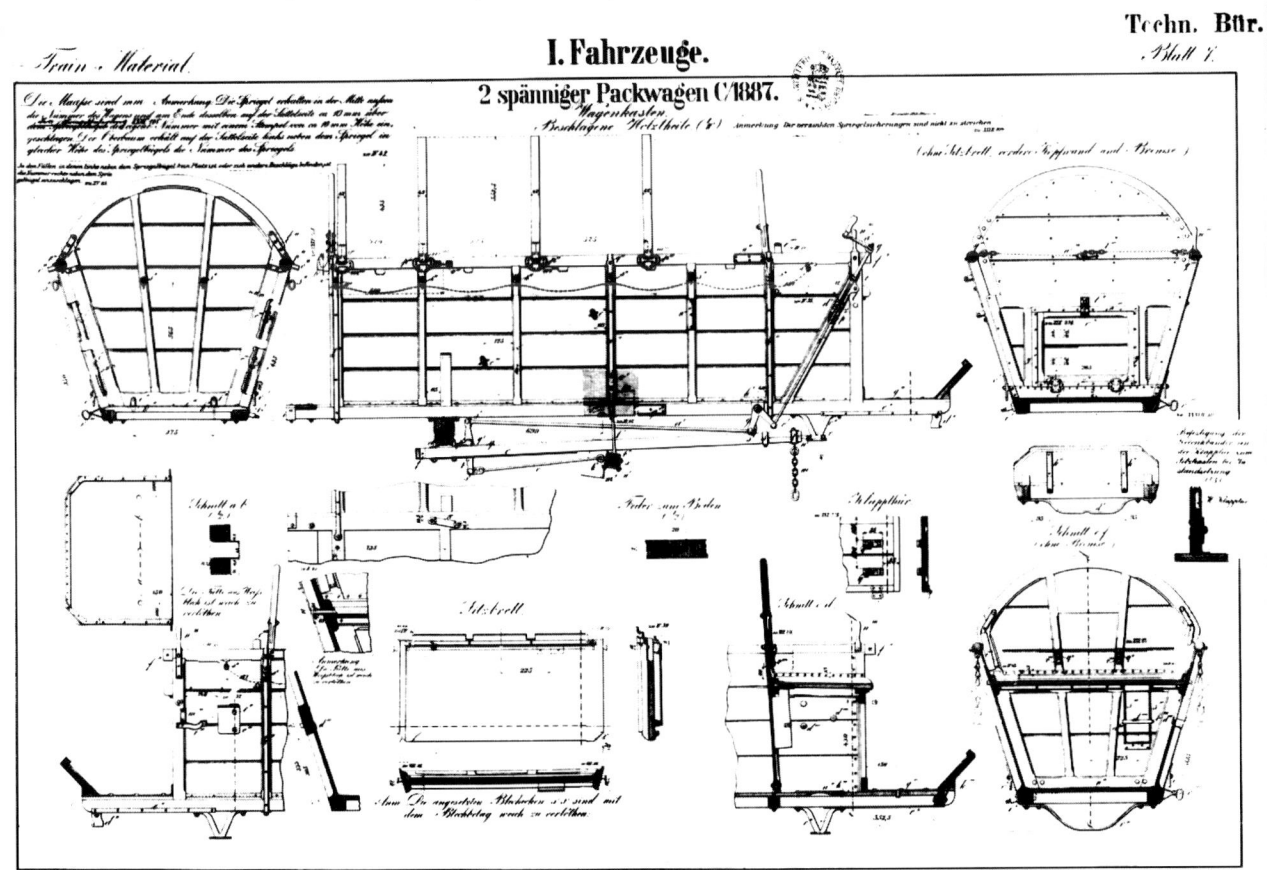

The Baggage Wagon 02 was used as the successor model by the heavy batteries of the heavy artillery. Pulled by two-horse hitches, it was used in the large baggage trains.

In the combat units of the heavy batteries equipped with 15 cm howitzers, eight, or sometimes twelve foot artillery ammunition wagons were used. This wagon type, designated Af. 5, was still used by the horsedrawn heavy batteries in World War II.

Because of the greater shot ranges, the heavy artillery required special observation wagons, in order to direct artillery fire more quickly. This picture shows Observation Wagon 94.

Medical companies used eight of the two-horse Medical Wagon c/87 (Ambulance c/87). This had a weight of 0.78 tons and could be used to transport four badly wounded men on stretchers.

To be able to make short-term repairs to weapons and equipment on the battlefield, supply wagons with spare parts and tools were used. This picture shows the supply wagon for heavy field howitzer batteries.

In World War I the Light Field Wagon 95 proved to be especially useful. It weighed 0.51 tons empty and, without its shaft, was 3695 mm long, 1530 mm wide, and 2120 mm high with its canvas top.

Another vehicle that was used more and more by baggage trains and ammunition columns in World War I was the Heavy Provision Wagon 05. It weighed 0.75 tons empty (measuring 4200 x 1850 x 2100 mm) and could carry up to 1,000 kg. Light ammunition columns had 17, heavy provision columns 27 four-horse wagons of this type. The Wehrmacht used it with slight modifications as Heavy Army Field Wagon Hf. 2.

*The Military history Museum of the Bundeswehr in Dresden has a collection of over twenty field wagons and handcarts—including this superbly restored Heavy Provision Wagon 05, which was salvaged from a mill near Nossen.*

## Order No. 36 of January 23, 1908
## Field Gray Paint of Field Equipment

Vehicles of the troop and training field equipment, including medical vehicles, machine-gun equipment, and the field equipment of the transport troops are in the future when newly built or completely repainted to be painted field gray, likewise, all other objects of the aforementioned devices and equipment, which to date have been painted gray, blue, or green. Because of the conditions of the communications equipment, see the loading instructions of the communications medical depots.

Setting off individual fittings, rivets, chains, etc. in black is to be abolished. These are likewise to be painted field gray.

Back walls, fire chambers, and smokestacks of mobile field baking ovens are to be painted in graphite as before. Supplies of older paint colors are to be consumed in touching up.

Sixt von Arnim

*Mobile field kitchens were first introduced in two brigades of the field army in 1908. Their development was based on a suggestion from the Magirus firm. By the time World War I began, some 1,100 of the two-horse field kitchens had been delivered to the army.*

In the field bakery columns, mobile field baking ovens were still used before World War I. A column possessed twelve of these four-horse vehicles, and the two platoons could bake 23,000 loaves of bread within 24 hours. This was half the daily needs of an army corps. The vehicle shown in the picture can be seen in the Military History Museum of the Bundeswehr in Dresden.

A contemporary instruction sheet showing the early version of the mobile baking oven. The wheels had been taken from the Cavalro Bridge Wagon 05.

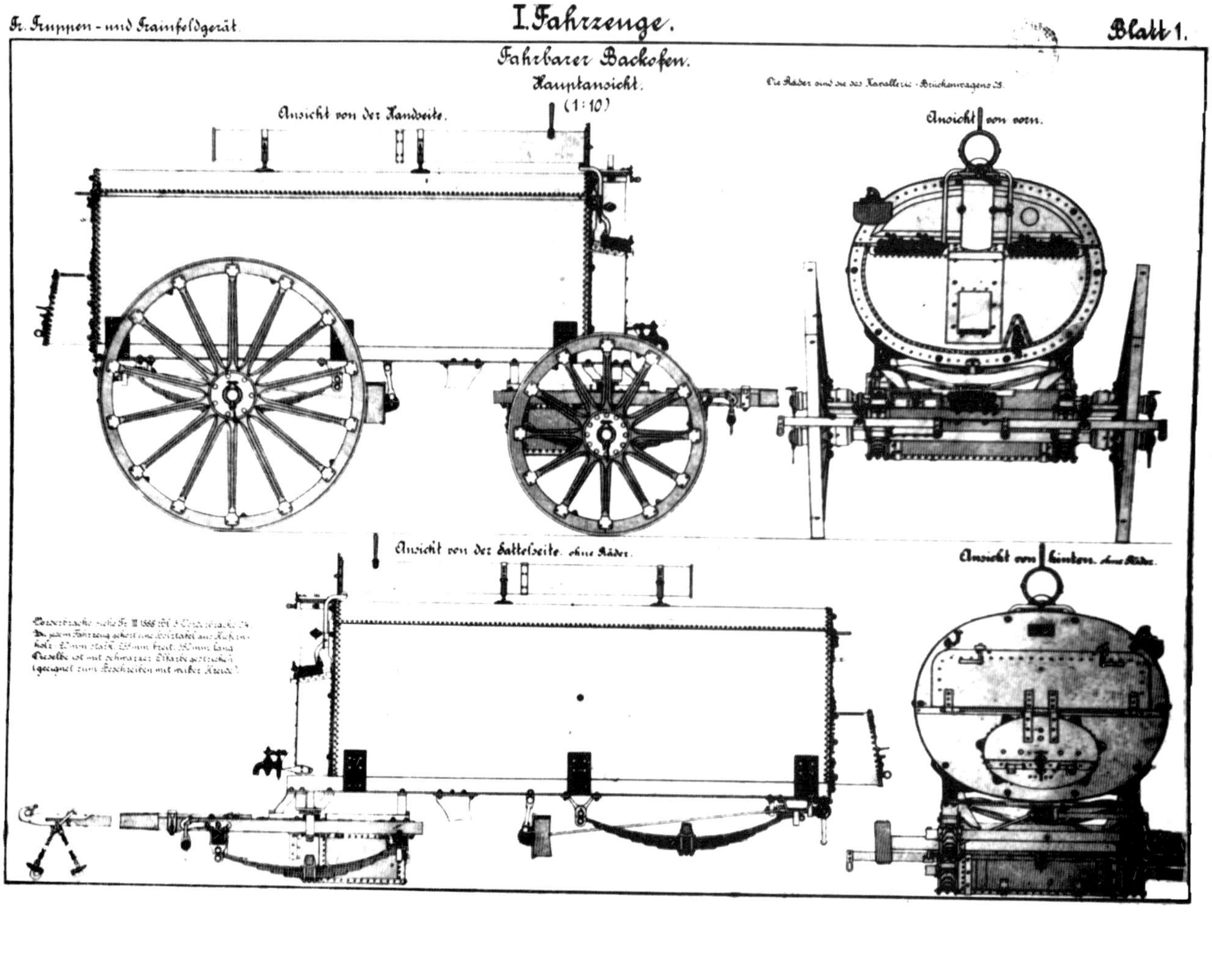

*Above: A field bakery column at work, photographed during the war in 1916.*


*Below: A train column in action during a prewar maneuver.*

*A properly loaded horsedrawn machine-gun wagon of the machine-gun company of the 11th Infantry Regiment No. 139 in Dobeln, photographed in pre-World War I times.*

*Vehicles of a machine-gun company on the march, photographed in the war years of 1915-1916.*

The medical care of soldiers played a major role in World War I, not only because of the increasing losses in the combat itself, but also because of the increasing danger of epidemics. The ambulance seen at right in the picture could carry four sick or wounded men.

Setting up a food supply column with 36 two-horse provision wagons, photographed on the eatern front in 1915.

*Vehicles of a field airship unit in a prewar maneuver.*

*Right: The wagon of a field radio station with its telescopic mast extended.*

*A teletype wagon of a telegraph unit.*

*Left: A limber for a 7.7 cm light mine launcher.*

*Handcarts for the transport of ammunition for the 7.7 cm light mine launcher were moved by manpower.*

*Handcarts for the transport of two mines for the 17 cm medium mine lanucher were also moved by manpower.*

*Next page, above: According to the terms of the Treaty of Versailles, the superfluous field vehicles that were not needed to equip the ten field divisions of the Reichswehr were to be destroyed.*

# 1919 - 1935

Germany came out of World War I as a loser. The stipulations of the Treaty of Versailles resulted in a reduction of the military forces, producing a considerably decreased military presence in the form of the Reichswehr.

In the course of reducing the fighting forces, the supplies of field vehicles were finally freed from the older, less practical, and sometimes damaged models. For example, the machine-gun wagons remained in use by the infantry, while the limbers, observation, and supply wagons were still used by the artillery. The situation was similar in the cavalry and engineers. Classified as a general army vehicle and still in use were the small and heavy field wagons, which had proved their worth in the war. The same applied to field kitchens, such as the Small Field Kitchen 17 that had been introduced during the war. Naturally, wartime experience was evaluated in the specialist press. For example, the "Militärische Rundschau" included a report that portrayed clearly what requirements must be made on heavy draft horses in the future. Practical changes were only made where the meager financial limits allowed. In May 1921 it was ordered that the Small Field Wagon 16 (Type 05) should be built new with a wheel design only 55 mm wide. As an entrenching-tool wagon, the Reichswehr still used the old Field Wagon 95. According to a directive of February 15, 1923, there were to be no special racks for entrenching tools in the future. They were to be "...kept in any boxes, even 88 mm shell crates." This was an attempt to keep expenses down.

The painting of field vehicles was also regulated anew in June 1922, whereby a clear distinction was made between those with multicolored (three-colored) paint and those painted in one color.

Only in the fiscal year of 1927 were the numerous cavalry regiments supplied with the large smithy wagon. It replaced the cavalry squadron smithy wagon that had originated during the war. Because of its meager carrying capacity it had not been very popular, and now it was to be used only for transport at the army bases.

Some units, including engineer columns, used the baggage wagon with field smithy, which had to be equipped by the troops themselves according to plans from the Army Weapons Office, instead of the large smithy wagon. Another attempt to save money.

Naturally, there were also efforts to create new field vehicles. One example: In the mid-twenties the Army Weapons Office tested field kitchens, made by the Senking Works of Hildesheim, at Kummersdorf.

Special attention was devoted to the mobility of the mine launchers that remained in the Reichswehr. There was a good reason: The artillery of the Reichswehr was only a torso; heavy artillery was completely absent. Thus, the 17 cm medium mine launcher was to be used as heavy artillery against close-range targets, and the 7.7 cm light mine launcher was to relieve the field artillery. They were used in the 13 (mine-launcher) companies. Mine launchers were drawn by two- or four-horse hitches. The same applied to the ammunition wagons made of front and rear sections.

As generally understood in our times, the former German Wehrmacht, which was built up energetically since the National Socialist takeover in January 1933, was a fully motorized and armored army. This is not accurate. It was true that great value was laid on hard-hitting and highly mobile armored divisions. In World War II they formed the main offensive power of the army and played a decisive role in all military successes. But it was also true that the German armament industry was not at all capable of thoroughly motorizing such a large fighting force as the Wehramcht represented, or of supplying it with the required fuels and lubricants. For that reason the larger part of the army consisted of partly motorized and horsedrawn divisions. This did not change very much in the war.

Thus, horses and field vehicles retained a considerable importance for the fighting forces and backline services. This concept included all the troops and services that

were needed to supply the field army. At first it was a matter of organizing the available supplies of vehicles and deciding what was practical. The new designs that appeared were based on models proved in World War I. But production of a number of vehicles made of steel was also prepared. All field vehicles were given numbers, grouped in the following categories:

| | |
|---|---|
| 1. Army vehicles for general use, general army equipment | = Hf. |
| 2. Army sleds | = Hs. |
| 3. Artillery vehicles | = Af. |
| 4. Infantry vehicles | = If., Itf. |
| 5. Engineer vehicles | = Pf. |
| 6. Vehicles of the information troops, communication vehicles | = Nf. |
| 7. Vehicles of the administrative troops, administrative vehicles | = Vwf. |

---

# 1936 - 1945

The following listing of field vehicles follows the above system's main groupings and includes wagons introduced during the war, but does not claim to be complete. The possible variety of designations, which was especially great for general-use army vehicles, can only be indicated with individual examples.

### Army Vehicles (Hf.)
Hf. 1 = Light army field wagon,
Hf. 2 = Heavy army field wagon,
Hf. 3 = Small army field wagon,
Hf. 4 - Mountain cart,
Hf. 7 = Steel field wagon,
Hf. 11 = Large field kitchen,
Hf. 12 = Small field kitchen,
Hf. 13 = Large field kitchen,
Hf. 14 = Small field kitchen;

### Artillery Vehicles (Af.)
Ammunition Wagon 96 n/A, Ammunition Rear Wagon 96 n/A,
Field Howitzer Limber 98. Ammunition Rear Wagon 98,
Ammunition Wagon 38 for 7.5 cm Field Cannon 38,

Field Cannon limber for 7.5 cm Field Cannon 38,
Light Field Howitzer Limber 98E,
Light Field Howitzer Limber 18 (horsedrawn),
Light Field Howitzer Limber 18/40,
Heavy Field Howitzer Ammunition Wagon,
Af.4 = Field Howitzer Ammunition Wagon,
Af.5 = Heavy Ammunition Wagon 02,
Af.7 = Barrel Cart for 10 cm Cannon 17,
Af.8 = Barrel Wagon for 15 cm Cannon 16,
Af.12 = Observation Wagon,
Af.19 = Barrel Wagon for Heavy 10 cm Cannon 10 and Heavy 15 cm Field Howitzer 10 (horsedrawn);

### Infantry Vehicles (If.)
Mountain Cart for one-horse M.15 for 3.7 cm Antitank Cannon M.37,
Grenade-launcher Cart with frame M.24/35 for 8 cm Grenade Launcher M.36,
Ammunition Cart for 8 cm Grenade Launcher M.36,
If.3 = Machine-gun Wagon, heavy (Type 08), front and rear wagons,
If.4 = Machine-gun Wagon, light (Type 08),
If.5 = Machine-gun Wagon (Type 36) for MG 34,
If.8 = Infantry Cart,
If.9 = Combat Cart for Heavy Grenade Launcher (8 cm),

If.12 = Ammunition Wagon,
If.12I = Limber for one 3.7 cm or 7.5 cm Antitank Gun,
If.13 = Mine-launcher Ammunition Wagon,
If.14 = Ammunition Wagon, front and rear wagons,
If.15 = Box Wagon, front and rear wagons;

### Engineer Vehicles (Pf.)
Engineer Equipment Wagon for Cavalry Staff Regiments
Pf.21 = Flamethrower Fuel Wagon,
Pf.22 = Engineer Handcart,
Pf.25 = Handcart for light Charge Launcher;

### Communication Vehicles (Nf.)
Construction Wagon 13,
Rack Wagon for Telephone Troops,
Box Wagon for Telephone Troops,
Nf.1 = Heavy Telephone Wagon 35,
Nf.2 = Small Telephone Wagon,
Nf.3 = Light Radio Wagon, front and rear wagons,
Nf.4 = Small Radio Wagon,
Nf.6 = Radio Wagon;

### Vehicles of the Administrative Troops (Vwf.)
Vwf.1 = Baking Oven Wagon,
Vwf.2 = Dough-kneading Wagon,
Vwf.3 = Power-source Wagon;

Let us look next at the general army vehicles, which were of much importance for all troops and service arms. They were used to transport all kinds of objects the army needed. The Hf.1 Light Field Wagon was drawn by two horses, weighed 0.61 tons empty, and was 3860 mm long without its shaft. The wheel diameter was 1224 mm, the tire width 55 mm. The following uses and versions of this wagon existed:

**1. In standard form:**
- Combat wagon,
- Ammunition wagon,
- Trenching-tool wagon,
- Equipment wagon,
- Provision wagon,
- Medical supply wagon,
- Veterinary supply wagon,
- Veterinary service wagon, and many more.

**2. Special types, sprung:**
Hf.1/1 = Communication equipment wagon,
Hf.1/1 = Engineering equipment wagon,
Hf.1/1 = Measuring equipment wagon,
Hf.1/2 = Searchlight wagon (M),
Hf.1/3 = Searchlight wagon (S).

**3. Special types, unsprung:**
Hf.1/1 1 = Large combat wagon,
Hf.1/13 = Large smithy wagon,
Hf.1/14 = Baggage wagon with field smithy,
Hf.1/14 = Weaponmaster's wagon,

Hf.1/15 = Baggage wagon for medical company,
Hf.1/16 = Bakery equipment wagon.

The Heavy Field Wagon Hf.2 had an empty weight of 0.8 tons and was made to carry loads up to 1.2 tons.

It was pulled by four horses (the average pulling power per horse was some 500 kg). Without its tongue the Hf.2 was 4250 mm long, the wheel diameter was 1224 mm, and the tire width 70 mm. It was used only in its basic form for the following purposes:
- Heavy combat wagon,
- Baggage wagon,
- Supply wagon,
- Provision wagon,
- Freight wagon,
- Ammunition wagon, and with modifications as a make-shift substitute for special vehicles.

Finally, the Hf.3 Small Field Wagon shall be portrayed. In its basic form it weighed 0.462 tons empty, could carry a load of 0.61 tons, and was 3250 mm long without its tongue. The wheel diameter was 1100 mm, the tire width 55 mm. The vehicle and its modifications were used for the following purposes:

**1. In its basic form**
- Ammunition wagon,
- Entrenching tool wagon,
- Equipment wagon,
- Repair shop wagon,
- Supply wagon,
- Baggage wagon,
- Provision wagon,
- Freight wagon, and with modifications for various special uses.

**2. Modified forms, sprung:**
Hf.3/1 = Communication equipment wagon,
Hf.3/2 = Light telephone wagon;

**3. Modified forms, unsprung:**
Hf.3/11 = Small combat wagon,
Hf.3/12 = small smithy wagon.

A comparison of the army's wagons in use before World War II with those from before World War I makes the development attained in barely forty years clear. From the former variety of vehicles, three basic types remained, and they could be suited to the needs and wishes of the troops. As before, they were made of wood. Only after the experiences of the Polish campaign, in September 1939, did the troops request that wagons be fitted with steel tongues. All wagons were built by the limber-nail system, with the tongues and other hitch parts (except in the small field wagon) uniform. The length of the crosstree of the light and heavy field wagons and the large field kitchen was 825 mm, those of the small field wagon and the small field kitchen 710 mm. The loose front member made of steel pipe with wooden fittings had a uniform length of 1400 mm.

*The four horses of a field howitzer ammunition wagon Af. 4 on the eastern front in the late summer of 1941 show what burdens the draft animals had been exposed to in the past months.*

The load areas of the three field wagons compare as follows:

| Wagon type | Space (cubic meters) |
|---|---|
| Light Field Wagon Hf.1 | 1.3 |
| Heavy Field Wagon Hf.2 | 2.0 |
| Small Field Wagon Hf.3 | 0.75 |

According to the II.Dv.476/1 of May 22, 1936, field wagons were painted three colors outside and field gray inside. This included the coloring of the canvas tops.

The older field wagons showed minor differences in their construction.

Even before World War II broke out, the first Hf.7 steel field wagons had reached the troops. The Hf.7, in a sense, marks the end of field-vehicle development in Germany. It differed essentially in its welded all-steel construction, four wheels with rubber tires, and swing axles with coli springs. By attaching a special towbar, the Hf.7 could also be used as a truck trailer. The empty weight was 1.04 tons, the load limit 1.5 tons. Without a tongue the wagon was 4103 mm long, 1810 mm wide, and 2100 mm high. Two horses pulled it. The towing limit for each animal was 1300 kg. Under the miserable conditions on the eastern front from 1941 on, this was definitely too much. The troops complained of this again and again. The Hf.7 was also put to use in various ways, such as the Large Combat Wagon Hf.7/11. The objects to be carried had to be loaded according to a loading plan. Outside the wagon was painted in camouflage colors, and inside field gray.

Among the typical field vehicles that must be regarded as general army equipment were the field kitchens. The Hf.12 small field kitchen weighed 0.72 tons, was 3410 mm long with its front wagon, and 1125 mm wide. The wheels and other parts were like those of the Hf.3 small field wagon.

Now let us look at another group of vehicles, the infantry wagons. Noteworthy here was the If.3 machine-gun wagon with twin mount 36, introduced in 1937.

The twin mount held two MG 34 guns set up for anti-aircraft use, and were meant particularly for providing marching infantry columns with protection from low-flying air attacks. The vehicles were used in the 4th (heavy) companies of the infantry battalions. To the extent that low-flying attack planes achieved greater speeds and became better armored, the two-horse Twin-Fla-MG (7.92 mm caliber) lost importance.

A different infantry vehicle, the If.8, owed its introduction to experience gained in the Blitzkrieg campaigns of 1939-40. It had been clear that the non-motorized units of the armored divisions could keep up only with the greatest effort. The industry was asked to provide marching performance that turned out to have a negative effect on their combat capability. A large portion of the ammunition and provisions had to be taken along, and carried on slow-moving requisitioned handcarts and wagons. A suitable vehicle of sheet steel was introduced as Infantry Cart If.8. On August 2, 1941, the D 193/1 "Infantry Carts, Their Use and Handling" manual came out. Various versions are known, with wooden-spoked wheels, and also with rubber and steel-tired disc wheels. The length was 1950 mm, the width 1000 mm, and the height 770 mm. Wheels with steel tires had a diameter of 670 mm.

The If.8 was moved by manpower, or by one horse or mule. Several wagons could be hitched one after another. The possible uses for transporting ammunition, provisions, fodder, weapons, and equipment were almost unlimited. The If.8 could be found everywhere: in the foglaying platoons of the staff companies of the mountain battalions (as double carts for twice ten 10 cm launch grenades in the ammunition units); in the corps machine-gun battalions established in the autumn of 1944 to transport, among others, radio sets and decoding machines; and to transport antitank rockets in the antitank units of the tank-destroyer brigades, to name only a few examples. In addition, the engineers used similar handcarts for uses that included transporting the Flame Thrower 41 or the light charge launcher plus ammunition.

The Russian campaign again created difficult conditions for all army field vehicles and their horses. On July 12, 1941, General Brand, coming back from a trip to the eastern front, already reported to the Army High Command on the rough road conditions and the stiff demands on the horses. Fodder shortages and diseases had resulted in many losses. Yet the horsedrawn columns took on greater importance in supplying the troops because of the extent of the front, the complicated road situations, and the great losses of motor vehicles, some through technical damage (dust effect). To make up for the losses of horses, large numbers of Slavic horses were used, which were tough and hardy, but not strong enough for the heavy German field vehicles. At times, local farm vehicles had to be used to get supplies to the front at all. On November 19, 1941, General Halder noted for a report to Hitler: "Condition of the division: very strained, horses!"

In September 1942, contracts were issued for 54,000 winter runners for field vehicles, in order to make them usable in winter. Lessons had been learned from the bad experiences of the winter of 1941-42. Such runners, some cobbled up by the troops themselves, were also used on guns. In addition, large numbers of army sleds were sent to the troops. The Hs.1 was also used with a wooden body as a medical sled. Others were used as platforms for light guns and grenade launchers.

The production of infantry carts, field wagons, and bicycles increased strikingly during the war. For 1943 and 1944 the following figures apply:

|  | 1943 | 1944 |
| --- | --- | --- |
| Infantry Carts | 38,500 | 40,700 |
| Field Wagons | 120,400 | 150,000 |
| Bicycles | 689,135 | 513,355 |

Yet, the lack of field vehicles forced shortages in equipping divisions. A 1945 type infantry division was supposed to have 1,273 field vehicles. For the newly established people's grenadier divisions, Hf.1, Hf.2, and Hf.3 were intended, as well as the Ersatz Field Wagons Erfa.40 and Ef.43, which were used with four horses to carry ammunition, plus the Pleskau 1 and 2 farm wagons. A fortress machine-gun company was also supposed to be equipped with vehicles of this kind in 1944. The lack of workers and materials made itself felt everywhere. In the Assault Program of the Army by the Army High Command/Army General Staff, Section III, of January 9, 1945, it was stated under Point 8, General Army Equipment: Equipment problems and extensive demotorization of the infantry compel a refinement and simplification of field vehicles, as well as ordered production and confiscation by the troops. Thus, the history of the German field vehicles came to an end. The end of the 19th century and the first half of the 20th, with their incredible variety of field wagons, can be regarded as the high point, but also as the end of the military use of horsedrawn vehicles. At the end of this time period came the final replacement of the horse as a military draft animal, with motor vehicles, tanks, and armored vehicles taking its place.

*A Heavy Field Wagon Hf. 2 of the 169th Infantry Division, seen at the Military History Museum of the Bundeswehr in Dresden.*

# Army Vehicles (Hf.)

*Above and below: a Light Field Wagon Hf. 1, which was used as a combat wagon by the army's rifle companies. Every platoon had such a wagon, with another ready for use by the company. The Hf. 1 was pulled by two horses, weighed 0.61 tons empty, and could carry a load of 1,360 kg. It was 3860 mm long, with a track of 1530 mm.*

One of the many special uses of the Hf. 1 was as a bakery equipment wagon (Hf. 1/16). According to H.Dv. 488/1, field wagons were painted in camouflage colors outside and field gray inside.

A variation of the Hf. 1 was the Light Field Wagon Hf. 1/1 with spring suspension of the body.

There were also many possible uses for the Hf.2 heavy field wagon. It was often used as an ammunition wagon by the field artillery of horsedrawn infantry divisions. Its empty weight was 0.8 tons; when it was loaded, the horses had to pull a maximum of two tons. Without a tongue the length was 4250 mm, the track was 1530 mm, and the wheel diameter 1224 mm.

The August Voges Wagon Factory in Hannover-Linden appeared as the manufacturer of the Hf.2 heavy field wagon, which appears to have been a provision wagon of the 169th Infantry Division, which saw service in Finland from 1941 to 1945, and then with the 9th Army on the Oder in the last weeks of the war. The paint is field gray (inside and out), and a light blue emblem with an elk's head is stenciled on the left front and the rear.

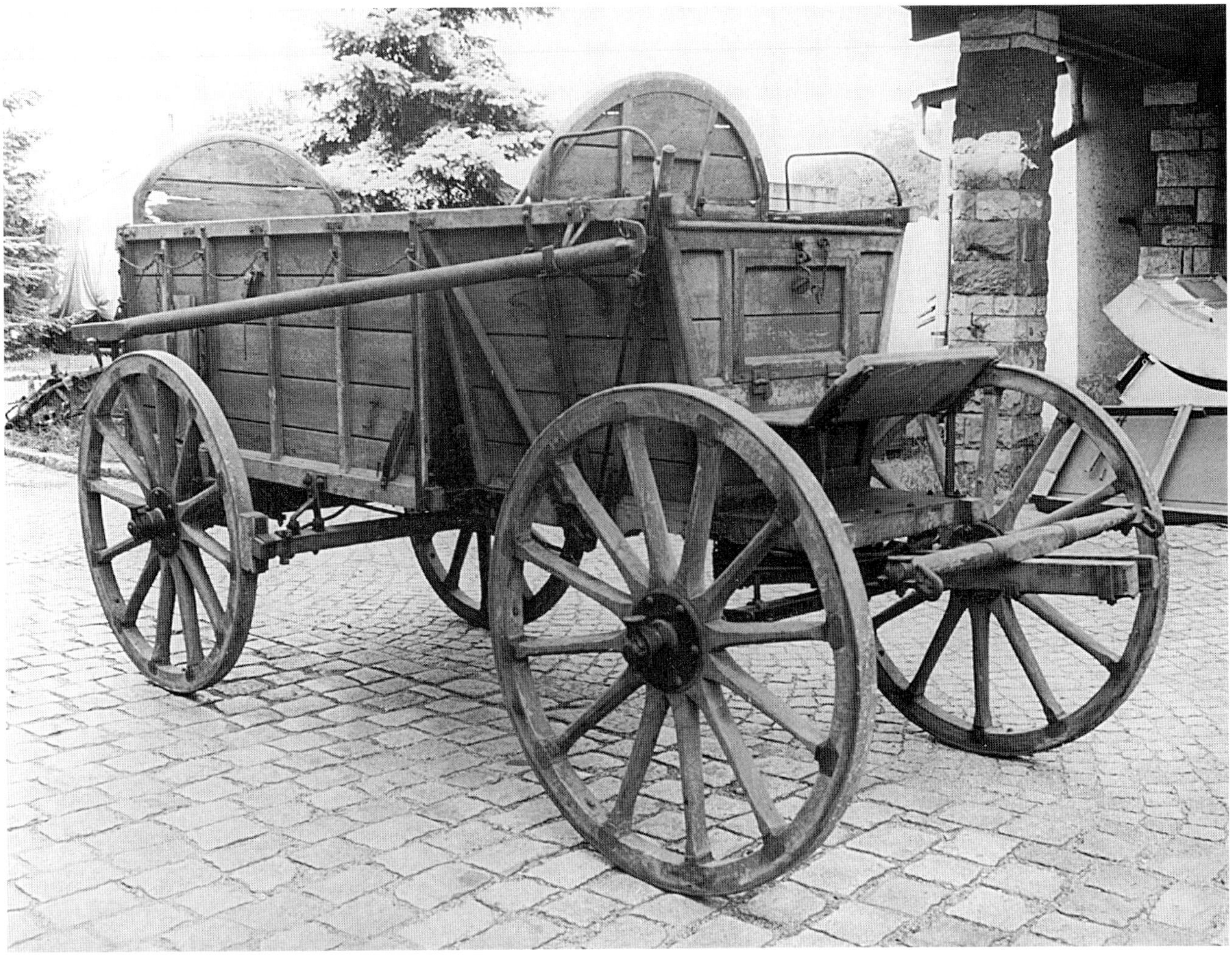

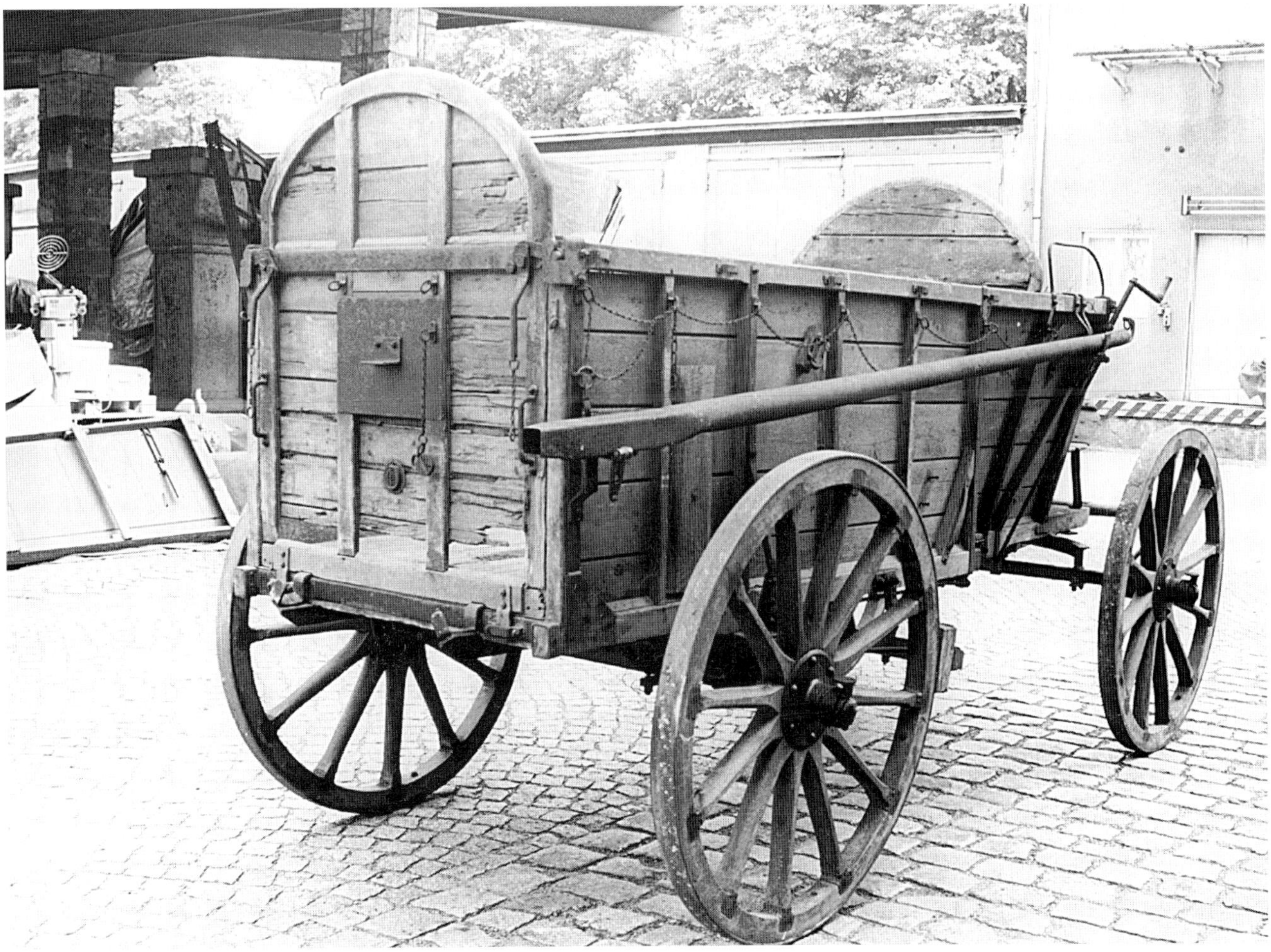

Supplying via muddy roads in the sector of the 11th Infantry Division near Sinjavino in the summer of 1943. The Hf.2 is loaded with balls of pressed fodder for the horses.

In the central sector of the eastern front in the summer of 1943. The Hf.2, weighing almost two tons, put a pressure of at least 540 kg on each horse. Thus, additional teams of horses had to help overcome rough terrain to spare the animals.

A team that was not foreseen in H.Dv. 465/1 of 1936: A column of Hf.2 heavy field wagons on the Don steppes in the summer of 1942.

The Hf.3 small field wagon came both sprung and unsprung. Typical versions were the small combat wagon (Hf.3/11) and small smithy wagon (Hf.3/12). In its basic form the wagon weighed 0.462 tons empty and 1.542 tons loaded, was 3250 mm long, and had a track of 1125 mm and a wheel diameter of 1100 mm.

Loading supplies from a truck onto an Hf.3 by a mountain unit in the autumn of 1942.

Oberer
Bockkasten

Unterer
Bockkasten

*The Hf.711 large combat wagon weighed 1.04 tons empty, and at most 2.76 tons loaded (including 70 kg of equipment and two drivers). Without a tongue it was 4103 mm long, with a track of 1580 mm and a wheel diameter of 835 mm. It could be hitched to a truck with a towbar.*

*The Sf.2 ambulance on tow by an 8-ton tractor (Sd.Kfz. 7) in the northern sector of the eastern front in the spring of 1942.*

*Another type of hitch not foreseen in the H.Dv. The average load pulled by each horse of a two-horse hitch was 1.38 tons. The first complaints were heard already during the Polish campaign in September 1939. Under the conditions faced in the Soviet Union, criticism escalated, and the Hf.7 got the reputation of being a "horse-murderer."*

In 1944-45 the very simplified Ersatz Field Wagons Erfa. 40 and Ef. 43 and the Pleskau 1 and 2 farm wagons reached the troops. The first strongly resembled typical farm wagons.

Captured horsedrawn vehicles, such as the French forage wagon of 1887 seen here, were put to use by the Wehrmacht. This picture was taken on the eastern front in June 1941.

A small field kitchen behind a tracked motorcycle (Sd. Kfz. 2) in North Africa in 1941.

A field kitchen of unknown origin, in use by a horsedrawn supply unit on the eastern front, seen in August 1941.

A small field kitchen of the III./Infantry Regiment 91 (27th Infantrry Division) in the winter of 1939-40. The kitchen weighed 0.72 tons; with a limber it was 3410 mm long and had a track of 1125 mm. The Hf.12 Small Field Kitchen was also used by communication units, engineers, and other troops.

An interesting and not unimportant detail of army life—the unit leader and the cook tasting the food. Will he like it? Photographed during prewar maneuvers.

# Artillery Vehicles (Af.)

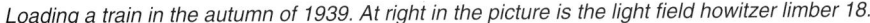

The light field howitzer limber with a 10.5 cm light field howitzer hooked on, photographed in France in 1940. The light howitzer limber could be used with a long steel tongue for horsedrawn use, and with a limber arm and coupling eye for motor transport.

Loading a train in the autumn of 1939. At right in the picture is the light field howitzer limber 18.

The Field Howitzer Limber 18/40 in this form was used by the light units of the artillery regiments that had been made mobile with the Raupenschlepper Ost (R.S.O.) tractor. The weight was 0.617 tons, the wheel diameter 1100 mm. This vehicle came from the Army Ordnance Department in Naumburg, and had been obtained by a farmer in the vicinity of Camburg after the war as a source of spare parts (wheels). In the mid-eighties, Military History Museum staff members salvaged two limbers out of a barn that was about to be dismantled. They still had the original dark yellow paint and stenciled black lettering.

The front of an Observation Wagon Af. 12, used in the batteries of the artillery units.

The Barrel Wagon Af. 19 was used by the horsedrawn cannon and howitzer batteries to transport barrels for the heavy 10 ch Cannon 18 and heavy 15 cm Howitzer 18. This one was photographed in Poland in 1939.

Below: The typical arrangement of a horsedrawn gun. Especially strong horses had been sought for the artillery. On the eastern front, shortage of fodder and heavy burdens soon brought on an acute shortage of such draft horses. Later, the Raupenschlepper Ost (R.S.O.) tractor took over the horses' tasks, at least in the light units.

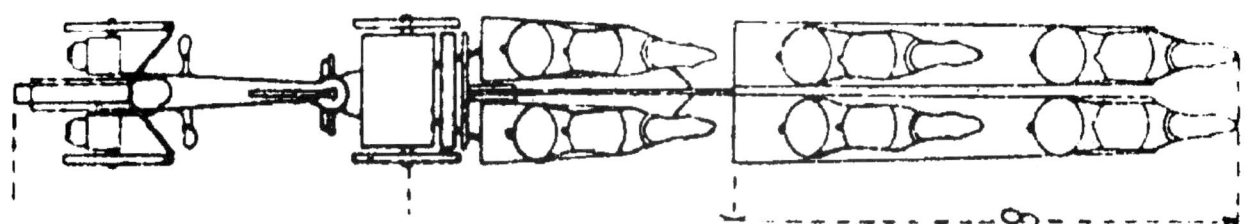

# Infantry Vehicles (If.)

*Mountain units used special mountain carts with mules to pull them—seen here carrying 2 cm Flak 30 guns during a victory parade in Athens.*

*The combat carts for the heavy grenade launcher, If. 9 and Ig.9/1, transported the 8 cm Grenade Launcher 34 and its ammunition (48 grenades). It had a forked tongue for one-horse use. This easily broke loose and gave cause for complaint during the Polish campaign in September 1939.*

The combat cart weighed 0.2 tons, was 3720 mm long, 1366 mm wide, and 1280 mm high, had a track of 1125 mm, a wheel diameter of 800 mm, and a body made of sheet steel.

A grenade-launcher platoon with six 8 cm Grenade Launcher 34 weapons used six lf.9 Combat Carts and three Hf.3/11 combat wagons.

The If. 3 Machine-gun Wagon (front and rear sections), each with two 08 Heavy Machine Guns, pass Hitler in a parade.

The horsedrawn MG companies and heavy MG half-platoons of the rifle companies carried their equipment on the If. 5 Machine-gun Wagon (Type 36). Pulled by two horses, it could carry two Maschinengewehr 34 guns and their mounts, equipment, and ammunition. The baggage of six gunners could be carried on the front wagon.

Vorderwagen

Hinterwagen

*The If. 5 Machine-gun Wagon (Type 36) carried twin guns to protect the infantry while on the march. The rear wagon could be braced by jacks with runners.*

*The Twin Mount 36 with two Machine Gun 34 weapons could be turned 360 degrees horizontally and from -10 to +90 degrees vertically. During the war the Machine Gun 42 was also used in place of the Machine Gun 34.*

An If. 5 Machine-gun Wagon (Type 36) of a horsedrawn infantry division in France in 1940.

The If. 8 Infantry Cart proved to be a very versatile vehicle during the war. Several of them could be used in tandem and drawn by one horse with a forked tongue.

Two If.8 Infantry Carts from the Military History Museum's collection. Along with these steel-tired wagons there were also those with spoked wooden wheels, plus those with steel disc wheels and rubber tires. Besides transporting supplies, they could be fitted with wooden bodies to carry the Grenade Launcher 34, Machine Gun 34 and 42, Panzerfaust and Antitank Rocket 54, as well as communications equipment.

The length without a tongue was 1190 mm, the width 990 mm, the height 730 mm; the empty weight was 81.5 kg, and a 350 kg load could be carried.

*Two 7.5 cm Light Infantry Gun 18 units hitched to limbers for the 7.5 cm Infantry Gun ltf. 12/1 (also used for the 3.7 cm antitank gun), with an lf.12 Ammunition Wagon at left.*

*Going against the original plans, which generally included motorized towing for antitank guns, the lack of suitable towing vehicles in 1942-43 caused horsedrawn infantry antitank units to be set up, using the 5 cm Antitank Gun 38 (seen here with the 11th Infantry Division in 1943) pulled by four or six horses, as well as the 3.7 cm antitank gun drawn by four horses behind the ltf. 12/1 limber.*

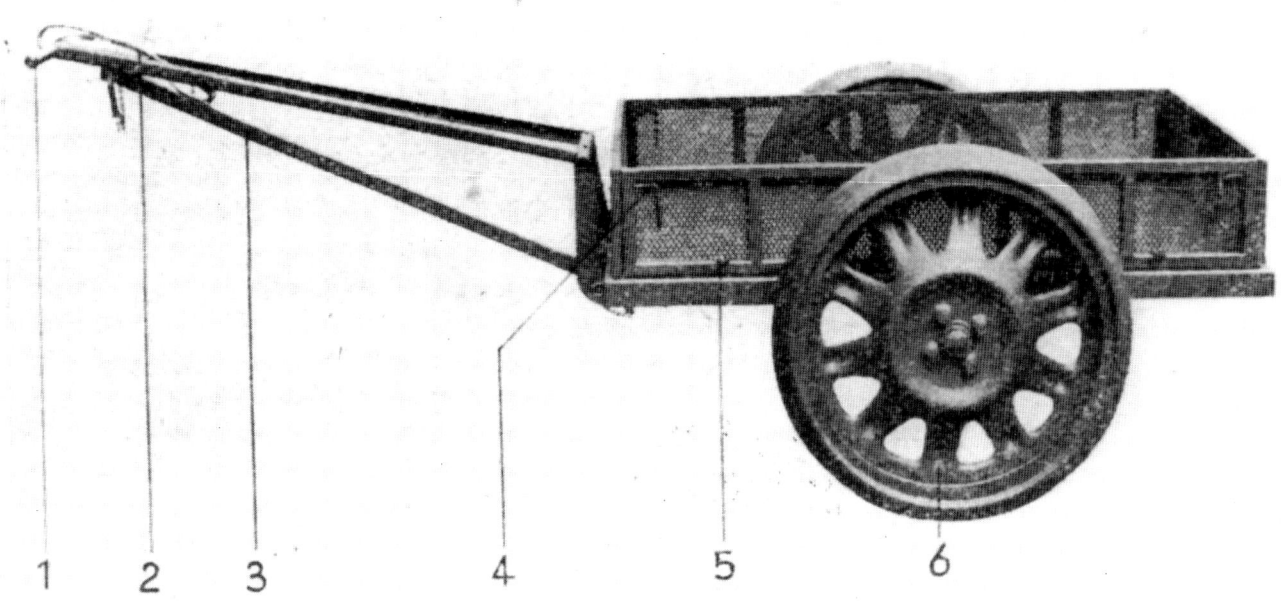

1    2    3              4         5         6

The Engineer Handcart Pf. 22 was carried on the engineer truck and could only be moved by manpower in action. It was used to transport flamethrowers, explosive and incendiary ammunition, small equipment, etc. It was 2460 mm long, 1140 mm wide, and 870 mm high. There were versions with pneumatic and with steel tires.

# Engineer Vehicles (Pf.)

A special weapon of the enginners was the light charge launcher. It could be dismantled and carried as three loads on the handcart for the light Pf. 25 Charge Launcher. This picture shows a Pf. 25 with the bottom plate of a launcher.

# Communications Vehicles (Nf.)

*Above: The Nf.7 Infantry Communications Wagon was used by rifle battalions; it differed slightly from the light telephone wagon.*

*Right and below: A radio station with Nf.3 Light Radio Wagon in operation, and an Nf.2 Radio Wagon of Artillery Regiment 4.*

# Army Sleds (Hs.)

*One army sled was this type, which was used to transport small guns (2 cm Flak, etc.) and other objects. A firm in Weixdorf, near Dresden, produced this sled early in 1945.*

*Boat sleds, or Bootskaya, were these light and low vehicles, which could be moved easily in unpacked snow. They could carry 150 kg, were 2370 mm long and 610 mm wide, and transported supplies, weapons, ammunition, or wounded men.*

*A Light Army Sled Hs. 1, capable of carrying 300 kg, from the Military History Museum of the Bundeswehr collection in Dresden. With a tongue it was 5200 mm long (without 282 mm), 1340 mm wide, and 900 mm high. There were also heavy army sleds that could carry 1,000 kilograms.*